Estimation of Multiple, Time-Varying Motions Using Time-Frequency Representations and Moving-Objects

S. Kannadhasan

M. Shanmuganantham

R. Nagarajan

R. Guruprasath

ELIVA PRESS

ELIVA PRESS

S. Kannadhasan
M. Shanmuganantham
R. Nagarajan
R. Guruprasath

A Wireless Sensor Network (WSN) comprises a collection of sensor nodes networked for applications like surveillance, battlefield, monitoring of habitat, etc. Nodes in a WSN are usually highly energy-constrained and expected to operate for long periods from limited on-board energy reserves. When a node transmits data to a destination node the data is overheard by the nodes that are in the coverage range of the transmitting node or the forwarding node. Due to this, the individual nodes might waste their energy in sensing data that are not destined to it and as a result the drain in the energy of the node is more resulting in much reduced network life time. As power is a limiting factor in a WSN, the major challenge in deploying a WSN is to enhance the network life time. So, it becomes inevitable to devise an efficient method of conserving the power. In this paper, a novel algorithm, Signed Graph based Adaptive Transmission Power (SGATP) is developed to avoid redundancy in sensing the data thereby enhancing the life time of the network. The concept of adapting the transmission power based on the distance of the next neighbor is proposed while a node communicates with the Cluster Head during Intrusion Detection. The simulation results show that the network life time is greatly improvised by the proposed method.

Published: Eliva Press SRL

Address: MD-2060, bd.Cuza-Voda, 1/4, of. 21 Chişinău, Republica Moldova

Email: info@elivapress.com

Website: www.elivapress.com

ISBN: 978-1-952751-91-2

Estimation of Multiple, Time-Varying Motions Using Time-Frequency Representations and Moving-Objects Segmentation

S. Kannadhasan

B.E.,M.E.,M.B.A.,PGDCA.,PGVLSI.,PGDRD.,MISTE.,MIE.,MIETE.,PGEMD.,
M.A.,M.Sc.,PGDBI.,(Ph.D)

Assistant Professor

Department of Electronics and Communication Engineering

Cheran College of Engineering

Karur, Tamilnadu-639111

M. Shanmuganantham

B.E., M.I.S.T.E

Lecturer (S.G) and Vice Principal

Department of Electrical and Electronics Engineering

Engineering Tamilnadu Polytechnic College Madurai-625011

Dr.R.Nagarajan

B.E., M.E., Ph.D

Professor

Gnanamani College of Technology

Namakkal-637018

R.Guruprasath

B.E., M.E.,

Assistant Professor

Gnanamani College of Technology

Namakkal-637018

OVERVIEW

We extend existing spatiotemporal approaches to handle time-varying motions estimation of multiple objects. It is shown that multiple, time-varying motions estimation is equivalent to the instantaneous frequency estimation of superpositioned FM sinusoids.

We apply established signal processing tools, such as time-frequency representations to show that for each time instant, the energy is concentrated along planes in the 3-D space: spatial frequencies—instantaneous frequency.

Using fuzzy C-planes, we estimate indirectly the instantaneous velocities. Furthermore, adapting existing approaches to our problem, we attain the identification of the moving objects. The experimental results verify the effectiveness of our methodology.

CONTENTS

CHAPTER 1 - INTRODUCTION

1. INTRODUCTION

The term digital image refers to processing of a two dimensional picture by a digital computer. In a broader context, it implies digital processing of any two dimensional data. A digital image is an array of real or complex numbers represented by a finite number of bits. An image given in the form of a transparency, slide, photograph or an X-ray is first digitized and stored as a matrix of binary digits in computer memory. This digitized image can then be processed and/or displayed on a high-resolution television monitor. For display, the image is stored in a rapid-access buffer memory, which refreshes the monitor at a rate of 25 frames per second to produce a visually continuous display.

1.1 THE IMAGE PROCESSING SYSTEM

A typical digital image processing system is given in Fig.1.1

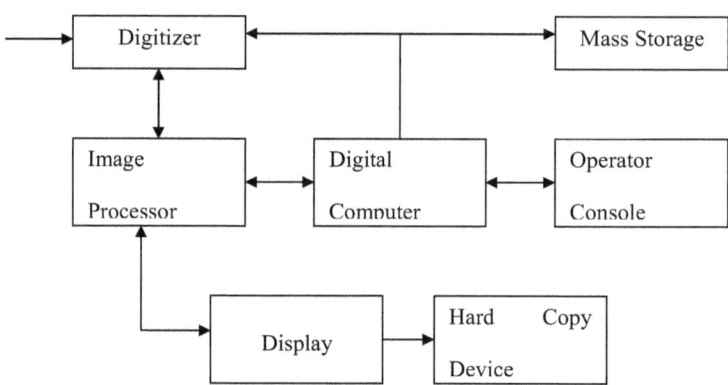

Fig 1.1   Block Diagram of a Typical Image Processing System

1.1.1 DIGITIZER

A digitizer converts an image into a numerical representation suitable for input into a digital computer. Some common digitizers are Microdensitometer.

1. Flying spot scanner
2. Image dissector

1

3. Vidicon camera

4. Photosensitive solid- state arrays.

1.1.2 IMAGE PROCESSOR

An image processor does the functions of image acquisition, storage, preprocessing, segmentation, representation, recognition and interpretation and finally displays or records the resulting image. The following block diagram gives the fundamental sequence involved in an image processing system

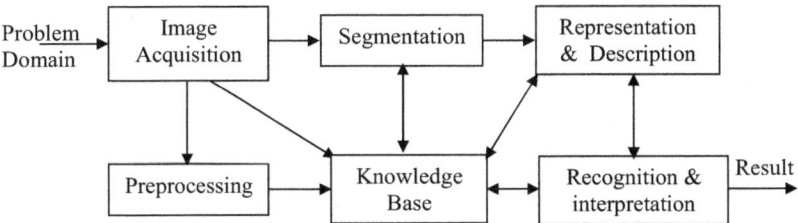

Fig 1.1.2 Block Diagram of Fundamental Sequence involved in an image Processing system

As detailed in the diagram, the first step in the process is image acquisition by an imaging sensor in conjunction with a digitizer to digitize the image. The next step is the preprocessing step where the image is improved being fed as an input to the other processes. Preprocessing typically deals with enhancing, removing noise, isolating regions, etc. Segmentation partitions an image into its constituent parts or objects. The output of segmentation is usually raw pixel data, which consists of either the boundary of the region or the pixels in the region themselves. Representation is the process of transforming the raw pixel data into a form useful for subsequent processing by the computer. Description deals with extracting features that are basic in differentiating one class of objects from another. Recognition assigns a label to an object based on the information provided by its descriptors. Interpretation involves assigning meaning to an ensemble of recognized objects. The knowledge about a problem domain is incorporated into the knowledge base. The knowledge base guides the operation of each processing module and also controls the interaction between the modules. Not all modules need be necessarily present for a specific function. The composition of the image processing system depends on its application.The frame rate of the image processor is normally around 25 frames/second.

2

1.1.3 DIGITAL COMPUTER

Mathematical processing of the digitized image such as convolution, averaging, addition, subtraction, etc. are done by the computer.

1.1.4 MASS STORAGE

The secondary storage devices normally used are floppy disks, CD ROMs etc.

1.1.5 HARD COPY DEVICE

The hard copy device is used to produce a permanent copy of the image and for the storage of the software involved.

1.1.6 OPERATOR CONSOLE

The operator console consists of equipment and arrangements for verification of intermediate results and for alterations in the software as and when require. The operator is also capable of checking for any resulting errors and for the entry of requisite data.

1.1.7 IMAGE

Image is the electronic form of picture. An image can be specified in a matrix form with each element of the matrix representing the basic unit of the image called pixel. A pixel is specified by allocation with its corresponding intensity denoted by f (x,y) where (x,y) represents the Spatial coordinates (position) and function f denotes the intensity. Based on the above parameters images are classified into three basic types.

- Binary Images
- Grayscale Images
- RGB Images

Binary images are those, which support only two colors namely black and white, their color values either 1 or 0. Gray scale images apart from black and white support 254 combinations of gray forming a total of 256 colors. RGB images support colors in eight or twenty four bit combinations leading to 256 or 16 million colors depending on the for most used. Images occupy memory when the size of the image increases or the number of colors over which it is operated increases. This causes problems like delay or distortion when an image is transported over a network. To overcome these compression techniques like GIF,JPEG and MPEG are followed depending on the type of image.

3

CHAPTER 2- DIGITAL IMAGE PROCESSING

2. DIGITAL IMAGE PROCESSING

2.1 INTRODUCTION

The term image refers to a two dimensional light intensity function, denoted by f(x,y), where the value or amplitude of f at spatial coordinates (x,y)gives the intensity of the image at that point. Light is a form of energy and is hence nonzero and finite. The images people perceive in day to day life consist of light reflected from objects. The basic nature of the image may be characterized by two components

   a. The amount of source light being incident on the scene being viewed and

   b. The amount of light reflected by the objects in the scene.

2.2 GRAY SCALE

The intensity of a monochrome image f at coordinates (x,y) is known as the gray level (l) of the image at that point. It is evident that

$$L_{min} \leq l \leq L_{max} \qquad \text{-------- (2.1)}$$

where,

L_{min} is the minimum gray level

L_{max} is the maximum gray level

And the only requirement is that L_{min} be positive and L_{max} be finite.

If i_{min} and i_{max} are the minimum and maximum values of the illumination and r_{min} and r_{max} are the minimum and maximum values of reflectance respectively, we have

$$L_{min} = i_{min} r_{min} \qquad \text{-------- (2.2)}$$

$$L_{max} = i_{max} r_{max} \qquad \text{-------- (2.3)}$$

4

The interval $[L_{min}, L_{max}]$ is called the gray scale of the image. Normally the image is shifted to the interval [0,L] where l=0 is considered black and l=L is considered white.

2.3 CLASSES IN IMAGE PROCESSING

An image processing system may handle a number of problems and have a number of applications but it mostly involves the following processes known as the basic classes in image processing

1. Image Representation and Description
2. Image Enhancement
3. Image Restoration
4. Image Recognition and Interpretation
5. Image Segmentation
6. Image Reconstruction
7. Image Data Compression

2.3.1 IMAGE REPRESENTATION AND DESCRIPTION

Any processed image must be represented and described in a form suitable for further computer processing. Basically, representing a region involves two choices

1. In terms of its external characteristics (its boundary) and
2. in terms of its internal characteristics (the pixels comprising the region)

The next task is to describe the region based on the chosen representation. Generally an external representation is chosen when the primary focus is on shape characteristics. An internal representation is selected when the primary focus is on reflectivity characteristics such as colour and texture.

Some of the available representation approaches are

1. Chain codes

2. Polygonal approximations

3. Signatures

4. Boundary segments

2.3.2 IMAGE ENHANCEMENT

The principle objective of enhancement technique is to process an image so that the result is more suitable than the original than for a specific application. Most enhancement techniques are very much problem oriented and hence enhancement for one problem may turn out to be degradation for the other. Enhancement approaches may be classified into two broad categories.

1. Spatial domain enhancement techniques
2. Frequency domain enhancement techniques.

The former refers to processing the image in the image plane (pixels) itself while the latter techniques are based on modifying the Fourier (or any other) transform of an image. In general enhancement techniques for problems involve various combinations of methods from both the categories.

Some examples of enhancement operations are edge enhancement, pseudo coloring, histogram equalization, noise filtering, unsharp masking, sharpening, magnifying, etc. The enhancement process does not increase the inherent information content present in the image but only tries to present it in a suitable manner. Enhancement operations may be either local or global.

2.3.3 IMAGE RESTORATION

The ultimate goal of restoration techniques (as in image enhancement) is to improve the image in some sense. However, unlike enhancement, restoration is a process that attempts to recover an image that has been degraded by using some apriori knowledge of the degradation phenomenon. Thus restoration techniques are oriented towards modeling the degradation and applying the inverse process in order to recover the original image. This approach usually involves formulating a criterion of goodness that will yield some optimal estimate of the desired result.

Early techniques for digital image restoration were derived mostly from frequency domain concepts. However, modern methods take advantage of the algebraic approach. Although a direct solution by algebraic methods generally involves the manipulation of large systems of simultaneous equations, under certain conditions computational complexities can be reduced to

6

the same level as required by traditional frequency domain restoration techniques. Restoration techniques may be either linear or non-linear.

Image restoration may be classified into three major types.

1. Restoration models: Image formation, detector and recorder, noise model, sampled observation.

2. Linear filtering Inverse / pseudo-inverse filter, Wiener filter, FIR filter, Kalman filter, semi recursive filter.

3. Other methods Speckle noise reduction, maximum entropy restoration, Bayesian methods, blind deconvolution, etc.

2.3.4 IMAGE RECOGNITION AND INTERPRETATION

Image recognition or analysis is a process of discovering, identifying and understanding patterns that are relevant to the performance of an image based task. One of the principle goals of image analysis is to endow a machine with the capability to approximate similar to human beings. An automated image analysis system is capable of exhibiting various degrees of intelligence. Some of the associated characteristics are

1. The ability to extract pertinent information from a background of irrelevant details.

2. The capability to learn from examples and to generalize this knowledge.

3. The ability to make inferences from incomplete information.

Image analysis can be divided into three basic areas.

1. Low level processing which deals with functions requiring no intelligence

2. Intermediate level processing which deals with the task of extracting and characterizing components in an image resulting form a low level process and

3. High level processing which involves recognition and interpretation and is generally termed as intelligent cognition

The predominant concept underlying image interpretation methodologies is the effective organization and use of knowledge about a problem domain. Current techniques for image interpretation are mostly decision – theoretic methods, some of which are predicate logic, semantic networks, expert systems, etc.

2.3.5 IMAGE SEGMENTATION

Image segmentation is a technique for extracting information from a image. This is generally the first step in image analysis. Segmentation subdivides an image into its constituent parts or objects. The level to which this subdivision is carried depends on the problem being solved. Segmentation is stopped when the objects of interest in an application have been isolated.

In general, autonomous segmentation is one of the most difficult tasks in image processing. This step determines the eventual success or failure of the analysis. Effective segmentation rarely fails to lead to a successful solution. Segmentation algorithms for monochrome images generally are based on one of two basic properties of gray level values

1. Discontinuity
2. Similarity

In the first category, the approach is to partition an image based on abrupt changes in gray level. The principal areas of interest within this category are detection of isolated points and detection of lines and edges in an image. The principal approaches in the second category are based on thresholding, region growing, region splitting and region merging. The concept of segmenting an image based on discontinuity or similarity of the gray level value of its pixels is applicable to both static and dynamic images. In the latter cases, motion can be used as a powerful queue to improve the performance of segmentation algorithms.

2.3.6 IMAGE RECONSTRUCTION

An important problem in image processing is to reconstruct a cross section of an object from several images of its trans-axial projections. A projection is a shadow gram obtained by illuminating an object by penetrating radiation. Each horizontal lines shown in the figure is a one dimensional projection of the horizontal slice of the project. Each pixel on the projected image represents the total absorption of the radiation along its path from the source to the detector. By rotating the source detector assembly around the object, projection views for several

8

different angles can be obtained. Image systems that generate such slice views are called computerized tomography (CT) scanners. These reconstructions are of several types.

1. Transmission tomography
2. Reflection tomography
3. Emission tomography
4. Magnetic resonance imaging
5. Nuclear magnetic resonance imaging

If a three dimensional object is scanned by a parallel beam, then the entire three dimensional object can be reconstructed from a set of two dimensional slices, each of which can be reconstructed using several available algorithms.

2.3.7 IMAGE DATA COMPRESSION

An enormous amount of data is produced when a 2-D light intensity function is sampled and quantized to create a digital image. The amount of data generated may be so great that it results in impractical storage, processing and communication requirements.

Image compression addresses the problem of reducing the amount of data required to represent a digital image. The underlying basis of the reduction process is the removal of redundant data. This amounts to transforming a 2-D pixel array into a statistically uncorrelated data set. The transformation is applied prior to storage or transmission of image. Later the compressed image is decompressed to reconstruct the original image or an approximation to it. Initial focus in this field was on the development of methods for reducing video transmission bandwidth, a process called bandwidth compression.

Image compression is the natural technology for handling the increased spatial resolution of today's imaging sensors and evolving broadcast television standards. Applications of data compression are in broadcast television, remote sensing via satellite, military communications via aircraft, radar and sonar, teleconferencing, computer communications, facsimile transmission, document and medical imaging, hazardous waste control applications and the like.

Image data compression methods fall mainly into 3 categories.

1. Pixel coding

    a. Run length coding

    b. Bit plane coding

    c. PCM/Quantization

2. Predictive coding

    a. Delta modulation

    b. Line by line DPCM

    c. 2-D DPCM

3. Transform coding

    a. Zonal coding

    b. Threshold coding

    c. Multi dimensional techniques

2.3.8 APPLICATION OF DIGITAL IMAGE PROCESSING

Digital image processing has a broad spectrum of applications such as remote sensing, image storage and transmission for business applications, medical imaging, acoustic imaging, and automated inspection of industrial parts.

Images acquired by satellites are useful in tracking of earth resources, geographical mapping, and prediction of agricultural crops, urban growth, weather, flood and fire control. Space imaging applications include recognition and analysis of objects contained in images obtained from deep space-probe missions. There are also medical applications such as processing of X-Rays, Ultrasonic scanning, Magnetic Resonance Imaging, Nuclear Magnetic Resonance Imaging, etc.

In addition to the above mentioned applications, digital image processing is now being used in solving a wide variety of problems. Though unrelated, these problems commonly require methods capable of enhancing information for human interpretation and analysis. Image enhancement and restoration procedures are used to process degraded images of unrecoverable objects. Successful applications of image processing concepts are found in astronomy, defense, biology and industrial applications.

3. REQUIREMENT SPECIFICATION

3.1 HARDWARE SPECIFICATION

- 800MHZ Processor

- 20GB Hard Disc

- 128 MB RAM

- Supported Motherboard

- 15"SVGA Color Monitor

3.2 SOFTWARE SPECIFICATION

OPERATING SYSTEMS:

- ❖ Windows 9x / 2000 / XP

SOFTWARE REQUIREMENTS

- ❖ MATLAB 7.0

3.3 SOFTWARE OVERVIEW

3.3.1 ABOUT MATLAB

MATLAB is a high-performance language for technical computing. It integrates computation, visualization, and programming in an easy-to-use environment where problems and solutions are expressed in familiar mathematical notation.

- Math and Computation
- Algorithm development
- Modeling, Simulation and Prototyping
- Data analysis, Exploration and Visualization
- Scientific and Engineering Graphics
- Application development, including graphical user interface building

3.3.2 WHERE MATLAB STANDS FOR

- ❖ MATLAB is an interactive system whose basic data element is an array that does not require dimensioning. This allows you to solve many technical computing problems, especially those with matrix and vector formulations, in a fraction of the time it would take to write a program in a scalar no interactive language such as C or FORTRAN.
- ❖ The name MATLAB stands for matrix laboratory. Today, MATLAB uses software developed by the LAPACK and ARPACK projects, which Together represent the state-of-the-art in software for matrix computation.

4. DEFINING THE PROBLEM

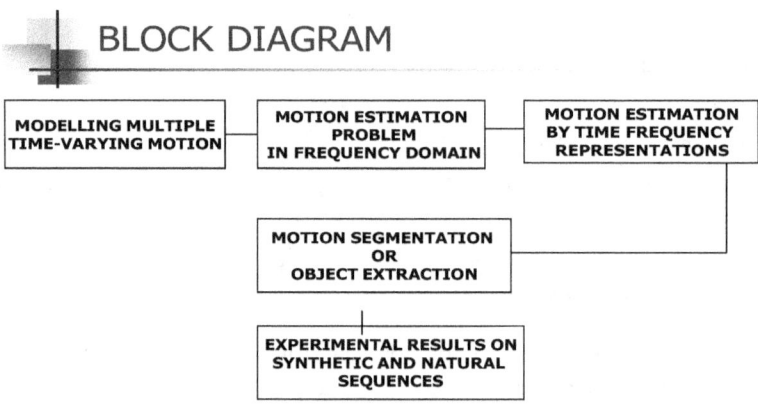

4. 1 MODELING MULTIPLE TIME-VARYING MOTIONS

We use a 2-D layered representation for image sequences. The layers are ordered in depth.
The layers are ordered in depth.

For each layer $1<=l<=L$,   We have

(a) Intensity map, $I_l(x)$

(b) Alpha map, $0<=a_l(x)<=1$

Where, 1=fully opaque

0=fully transparent

(c) Time-varying velocity $u_l(t)$

The layered representation can be summarized in the equation

$$f(x ; t) = I_l (x - d_l (t)) * w_l(x, t) \ldots\ldots\ldots\ldots\ldots\ldots\ldots..(1)$$

where, $d_l(t)=[d_{lx}(t), d_{ly}(t)]= u_l()d$, $w_l(x,t)=a_l(x-d_i(t))$ $(1-a_i(x-d_i(t)))$

Although the alpha map a_l remains registered with the intensity map I_l, this does not hold for w_1. The window $w1$ is deformed over time and has to be decomposed into a constant portion and a deformable, error-introducing portion. After some manipulations we get,

$$f(x ; t) = \sum_{l=1}^{L} f_l (x - d_l(t)) + e (x ; t) \ldots\ldots\ldots\ldots\ldots..(2)$$

4.2 VELOCITY ESTIMATION AS INSTANTANEOUS FREQUENCY ESTIMATION

Consider the spatial (2-D) Fourier transform of each frame.

From equation (2) with $w = (w_x , w_y)$ we have,

$$F(w ; t) = \sum_{l=1}^{L} F_l(w) e^{j \Phi_l(w ; t)} + E(w ; t) \ldots\ldots\ldots\ldots\ldots\ldots\ldots(3)$$

- Where , $w = (w_x , w_y)$ spatial frequency
- $\Phi_l(w ; t) =$ Instantaneous phase of lth component
- $\Phi_l(w ; t)= w_x d_{lx}(t) + w_y d_{ly}(t)$

    1) For a fixed spatial frequency w, the signal $F(w ; t)$ is a superposition of L frequency modulated (FM) complex sinusoids with instantaneous frequencies.

    2) For a given time instant t , the instantaneous frequency varies linearly with w_x, and w_y. For this purpose, one can exploit established signal processing   tools, such as the time-frequency representations.

14

4.3 TIME FREQUENCY IMAGES

The approach the instantenous frequencies to find the L peaks along v in the TFR image, for each t.However, in order to obtain robustness against noise and interference terms, and we use the following methodology. Considering that the instantaneous velocities and consequently the instantaneous frequencies v(t) as smooth functions of t, we expect smooth curves in the TFRplane, which can be approximated as piecewise linear polynomials.

Let the length of image sequence be T in the interval of [0,T] into N_p overlapping intervals of length T_p.

$$M_k : t \in (t_k - T_p/2, \ t_k + T_p/2) ,k=0,\ldots..N_p-1 \ldots\ldots\ldots\ldots\ldots\ldots..(4)$$

Where , T_p = length of time interval

T_k = center time instant of kth interval

N_p = No of overlapping time intervals

$$\downarrow \quad N_p = (\ T-T_p \ / \ T_p \ (1-\mu) \ ) +1 \ \ldots\ldots\ldots\ldots\ldots\ldots\ldots\ldots\ldots\ldots(5)$$

CHAPTER 5 – VIDEO INDEXING

5.1 VIDEO SUMMARISATION

A set of key frames that summarise the video content can be used in conjunction with existing textual annotations to augment the indexing process, to enable non-sequential browsing or to create a visual index into video that has not been previously textually annotated—as the saying goes "a picture is worth a thousand words". Selected key frames arranged as a storyboard can be used to quickly peruse the contents of a selected program. However, more importantly, if key frames are properly selected, many higher level content searches can be performed on the key frames rather than the complete video, thus reducing the computational requirements involved. The difficulty in composing a visual summary is determining which frames best represent the video contents and correctly portray the storyline.

Figure 5.1: Equally distributed frames representing a single shot - a single key frame would not represent all the shot's content.

To allow efficient indexing, a summary for a digital video library must represent the entire video content with as little redundancy as possible. Each key frame should represent a video segment which exhibits consistency in content. A common approach to this problem is to segment temporally a sequence into shots and then select a single representative key frame for each shot. The resulting ordered set of key frames is often referred to as a filmstrip. Although this method illustrates the practical use of segmenting a video into its constituent shots, one key frame may not always be sufficient to represent each shot. A shot can contain events such as camera or object motions that may drastically change its content. Fig. 5.1 shows 8 equally distributed frames from a single shot. At the beginning of the shot there is little activity, then two women appear from a building entrance and the camera pans right to track them walking down the street.

This would be difficult to represent using a single key frame, which leads to the idea of motion-based key frame detection. Motion analysis has been used previously to extract key frames. Wolf proposed a method that selects key frames at local minima of motion activity within a shot based on the assumption significant pauses are used to emphasise video content The author conjectures that in many shots key frames are identified by stillness—either the camera stops on a new position or the characters hold gestures to emphasise their importance .This method selects key frames which correspond to pauses in the video sequence between motion activity within a shot. Consequently, the content change between key frames can be expected to be the result of camera or object motion.

5.2 ESTIMATING THE DOMINANT MOTION

The first step in this method is to estimate the dominant motion between each consecutive frame pair. Given the 2D motion vector field obtained from the block-based motion compensation employed in the edit effect detection algorithm, a robust estimator is used to estimate the parameters of a simple motion model. Assuming the dominant motion between a frame pair is caused by camera motion, these estimates can then be used to identify shots containing significant camera motion that may require more than one key frame to represent their content.

The video segmentation algorithm employs hierarchical motion estimation to find the optimal block size to represent the visual content of each frame pair. The chosen block size is such that it provides the best overall correlation between the given frame pair. If there is little high frequency information and/or little motion between two frames a larger block size tended to be chosen.

If there exists multiple motions or motion that violates the 2D translational model between two frames a smaller block size was chosen. For this reason, we can assume the optimal block size to represent the visual contents also estimates the motion sufficiently accurately to estimate the dominant motion between a consecutive frame pair.

Common camera operations used in video production can be grouped into two broad classes: (i) tripod motion and (ii) free motion. If a camera is fixed to a tripod it can only exhibit three types of motion as shown in red in Fig. 3.2:

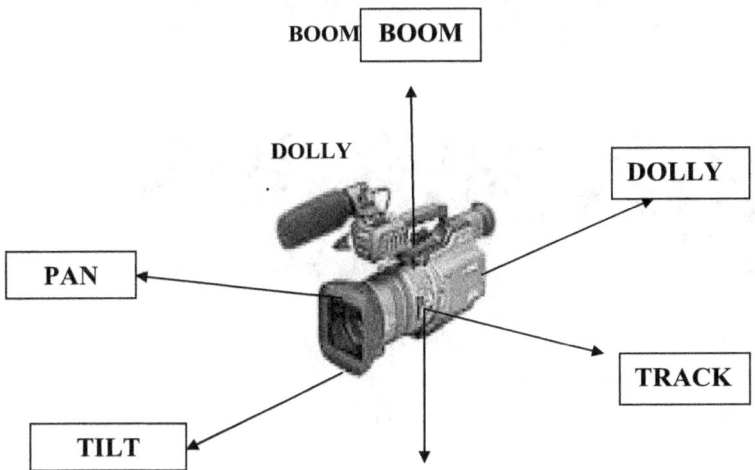

Figure 5.2: Common camera operations used in video production.

1. **Pan** - a rotational movement of the camera about the vertical axis.
2. **Tilt** - a rotational movement of the camera about the horizontal axis.
3. **Zoom** - convergent or divergent.

If there is free motion of the camera it can exhibit three additional motions shown in black in Fig. 5.2:

1. **Boom** - upward/downward motion of the camera along the vertical axis.
2. 2. **Track** - right/left motion of the camera along the horizontal axis.
3. 3. **Dolly** - forward/backward motion of the camera along the optical axis.

CHAPTER 6 -MOTION CHARACTERISATION

For each shot, the extracted key frames give a graphic, sequential depiction of the narrative; analogous to a storyboard. However, as well as browsing the visual content a user may also wish to perform a specific search. A search query could, for example:

1. Camera motion: pan, tilt, zoom etc.
2. Camera angle: high, low, aerial etc.
3. Camera position: close up, long shot, mid shot etc.
4. Lighting: artificial, daylight, dark etc.

CHAPTER 7- EXPERIMENTAL RESULTS

7. EXPERIMENTAL RESULTS

7.1 WHITE TOWER & BOAT SYNTHETIC SEQUENCE

The synthetically generated sequence o in the figure contains a moving object (boat) on a moving background (White Tower). The frame size is 296* 296 and the sequence's length is T=64 frames. The background moves with velocity

$V_{1x\ (t)}$=-1 pixel/frame and the boat decelerates according to $v_{2x(t)}$=3-0.05t pixels/frame. We used two 8-bit grayscale still images, a background "White Tower" image of size 502* 296 and an image containing the "Boat" object. The motion of the background was obtained by selecting the appropriate 296 *296 region of the corresponding still image, for each time instant. For the"Boat," we created manually an auxiliary "mask" image, which corresponds to the window. We cropped the "Boat" object from the still image and superimposed it on the

background frame, at the appropriate instantaneous location.

In order to handle non-integer instantaneous displacements, we applied bilinear interpolation. Finally, to obtain more natural visual appearance, we applied a weak smoothing operation near the border of the "Boat" object.

For this experiment, White Gaussian Noise with SNR=32dB was added. We use the spectrogram because of its simplicity. The window used to obtain the Short-Time Fourier Transform is a Hamming(T/4) . In the TFR images, we search for two peaks along v for each time instant t . This is equivalent to applying the proposed Hough-transform methodology on TFR slices of length T_p=1.

7.2 WALKER EXPERIMENT

The natural sequence in the figure contains a single walker, whose size is small compared to the static background. The sequence contains T=187 video-frames of size 320 *240. The TFR images are clear in the sense that almost no interference terms exist and the TF localization is good. Applying our methodology by cutting the TFR images into overlapping time-slices TFR of length T_p=T/18with overlapping factor μ=0.5 , the estimated instantaneous velocity However, our methodology is more general in the sense that it can handle also sequences with large moving objects and moving background, at the cost of increased computational effort.

7.3 TRANSPARENT LENNA-CAMERAMAN SYNTHETIC SEQUENCE

In order to verify the effectiveness of the approach for transparent motions, we created the synthetic sequence shown in the figure. The final sequence shown is of size 256* 256 and length T=32. To generate it, we used 8-bit grayscale, 512* 512 versions of the well-known "Lena " and "Cameraman" still images. The translating foreground"Lena " layer is opaque except from a together-translating rectangular region, where it is transparent.

According to the presented motion model, the corresponding "alpha" map is unity everywhere, except from the rectangular image region x,y $\in$ {50,...256}*{20,......200}(for the first frame), where it is equal to 0.5. To handle non-integer instantaneous displacements, we applied bilinear interpolation. White Gaussian noise (WGN) with SNR=23dB was added.

7.4 MODIFIED COAST GUARD EXPERIMENT

In this experiment, the well-known "Coast Guard" sequence is used. In order to have time-varying motion for the background, we have created a modified sequence in the length T=64 . The original sequence's frames 138–265 are used with an increasing rate, according to the formula: $n_1=138+n+0.015n^2$, with n and n_1 being the frame numbers in the new and the original sequence, respectively. However, applying the presented Hough-transform methodology, the estimation of the instantaneous frequencies becomes reliable.

The length of the slices is $T_p=T/4$ and the overlapping factor between successive pieces $\mu=0.75$.

7.5 CAVIAR SEQUENCE-MAN WALKING

The proposed methodology was successfully applied also on sequences captured for the purpose of the CAVIAR project. We present the results for the frames 310–437 of the original "Walk2" sequence captured at the INRIA-lab, with a time-down sampling factor equal to 2. The final sequence in the Figure has T=64 frames of size 384 *288 and contains a man walking along a parabolic trajectory. For the piecewise linear approximation of the instantaneous frequencies curves, the produced TFR images were cut into overlapping slices of length $T_p=T/8$, with overlapping factor $\mu=0.75$.

CHAPTER 8- FLOW CHART

8. FLOW CHART

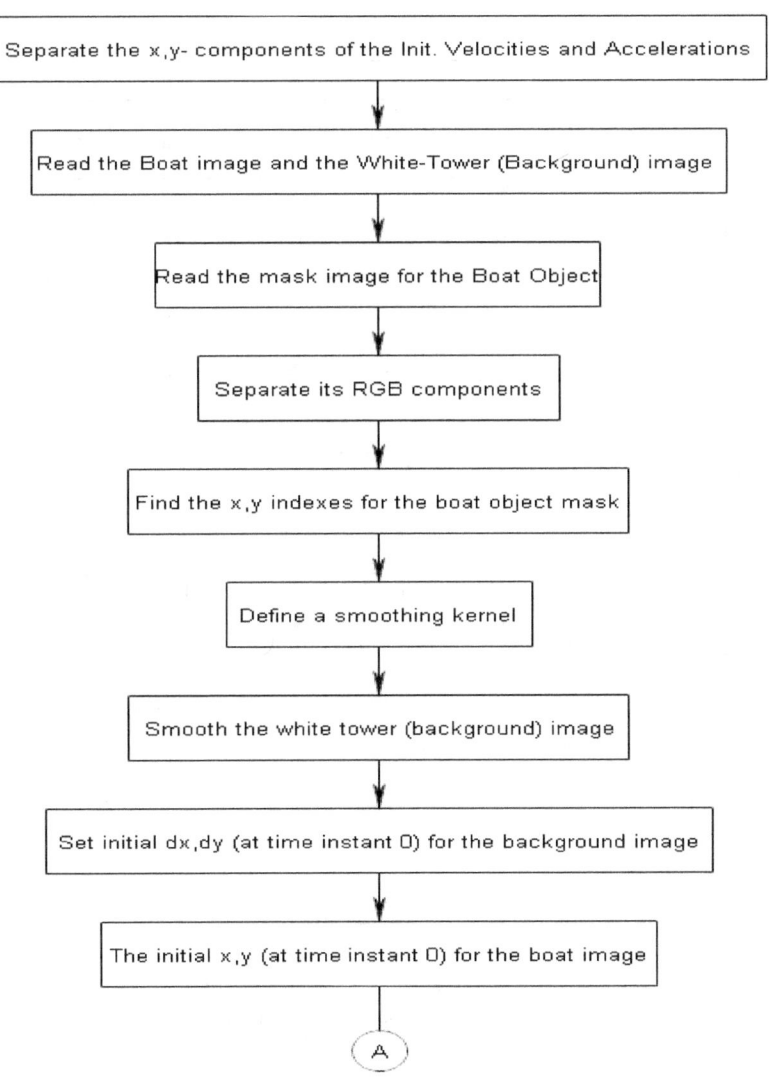

Separate the x,y- components of the Init. Velocities and Accelerations

Read the Boat image and the White-Tower (Background) image

Read the mask image for the Boat Object

Separate its RGB components

Find the x,y indexes for the boat object mask

Define a smoothing kernel

Smooth the white tower (background) image

Set initial dx,dy (at time instant 0) for the background image

The initial x,y (at time instant 0) for the boat image

A

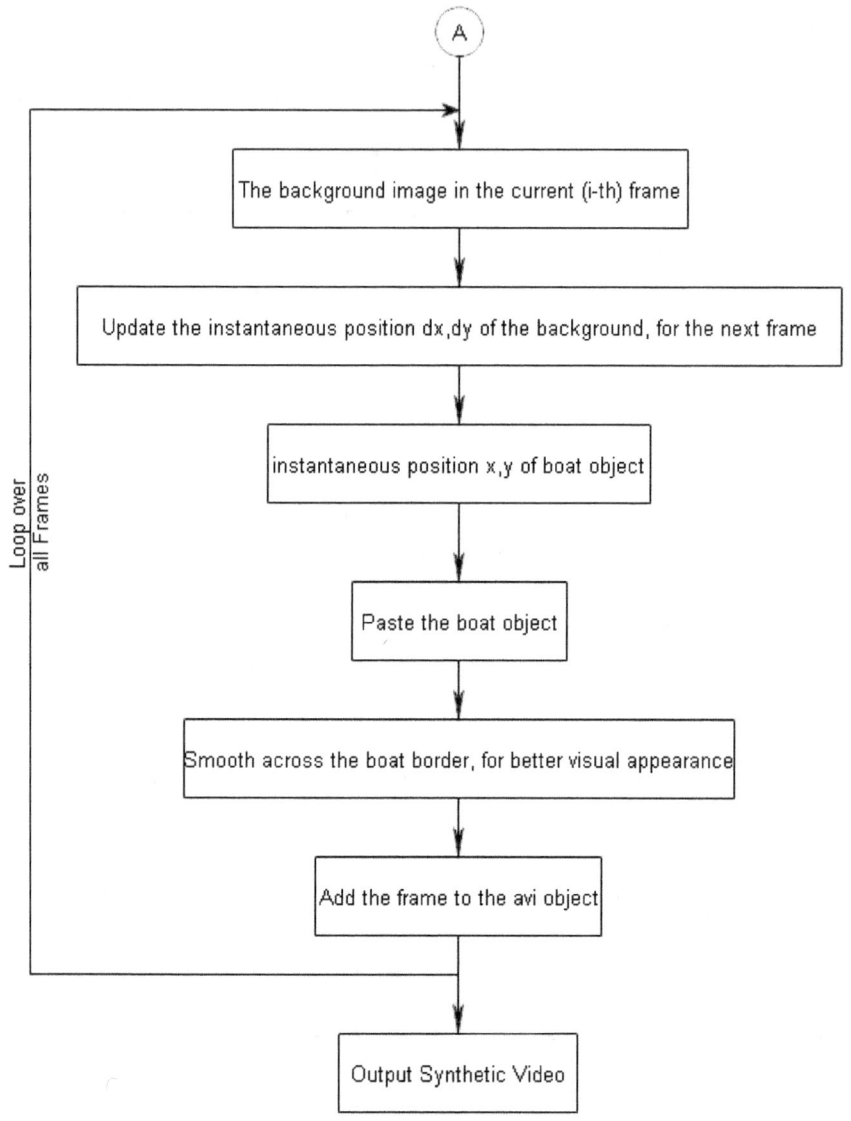

CHAPTER 9- RESULTS AND ANALYSIS

9. WHITE TOWER - BOAT EXPERIMENT

1.1 GRAPHICAL REPRESENTATION

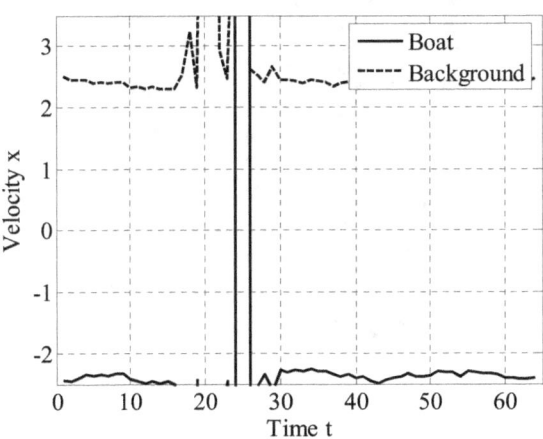

Figure 1.1(a)

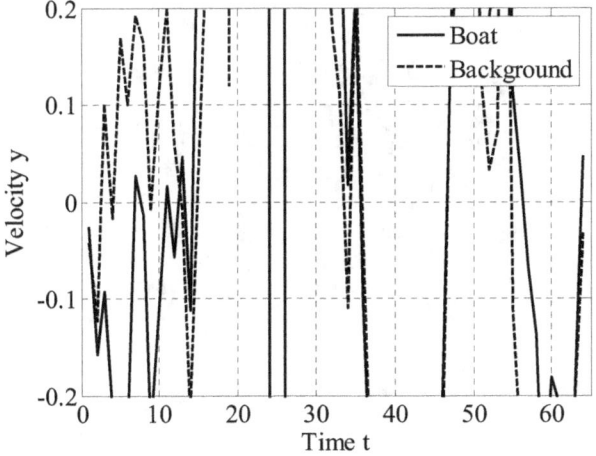

Figure 1.1(b)

1.2 PICTORIAL REPRSENTATION

Figure 1.2(a)

Figure 1.2(b)

2. WALKER EXPERIMENT

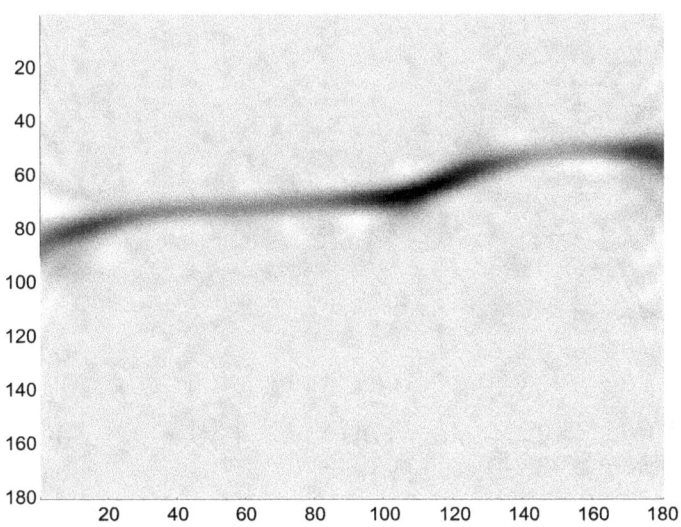

Figure 2(a)

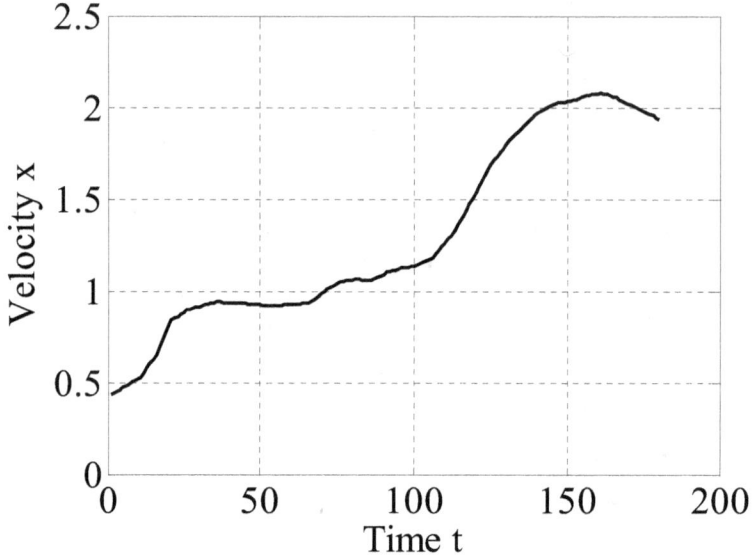

Figure 2(b)

27

3.TRANSPARENT LENNA - CAMERAMAN EXPERIMENT

3.1 GRAPHICAL REPRESENTATION

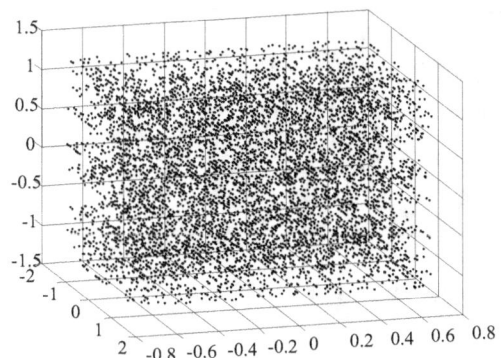

Figure 3.1(a)

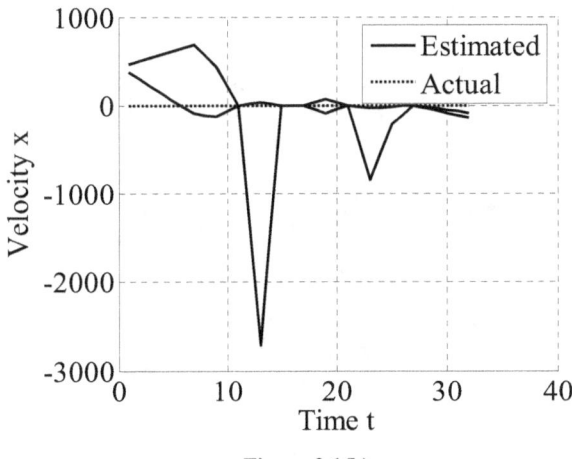

Figure 3.1(b)

3.2 PICTORIAL REPRESENTATION

Figure 3.2(a)

Figure 3.2(b)

4. COAST GUARD EXPERIMENT

4.1 GRAPHICAL REPRESENTATION

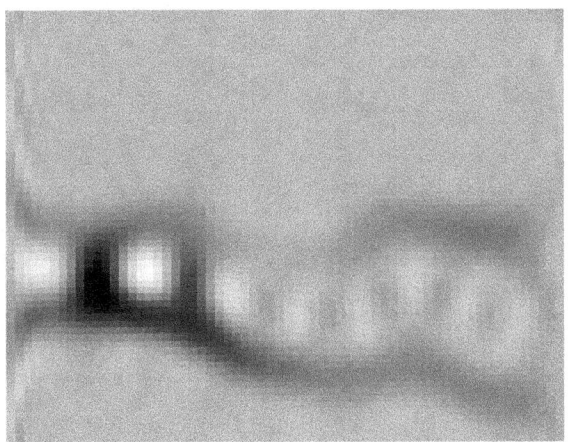

Figure 4.1(a)

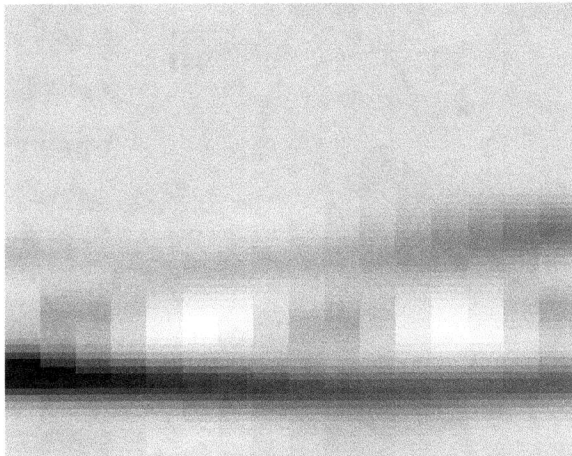

Figure 4.1(b)

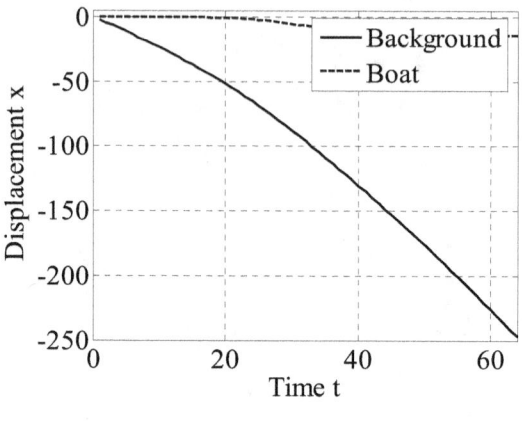

Figure 4.1(c)

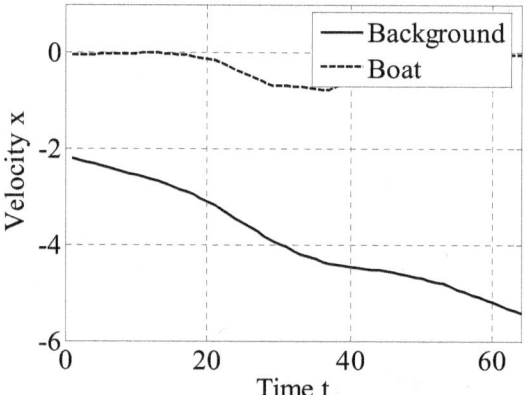

Figure 4.1(d)

4.2 PICTORIAL REPRESENTATION

Figure 4.2(a)

Figure 4.2(b)

5. CAVIAR EXPERIMENT
5.1 GRAPHICAL REPRESENTATION

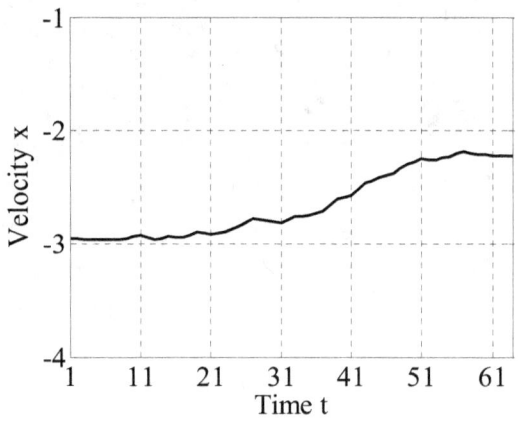

Figure 5.1(a)

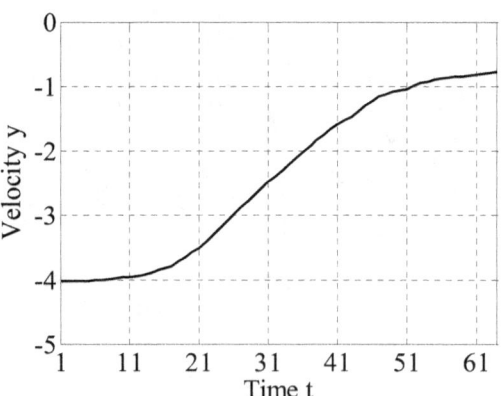

Figure 5.1(b)

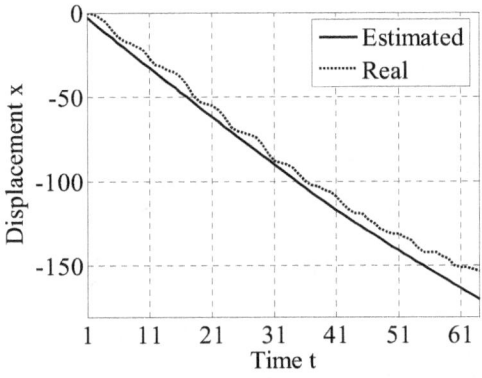

Figure 5.1(c)

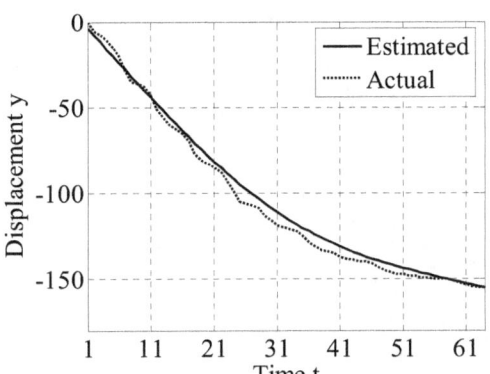

Figure 5.1(d)

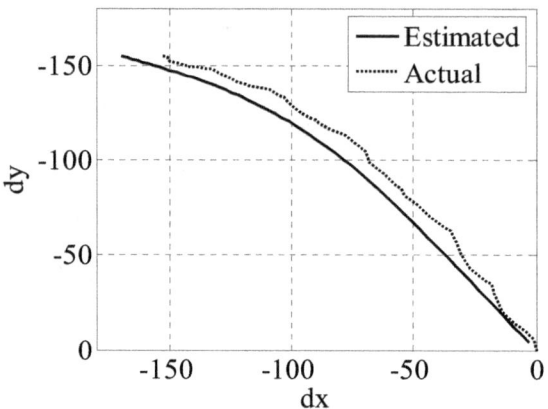

Figure 5.5(e)

5.2   PICTORIAL REPRESENTATION

Figure 5.2(a)

Figure 5.2(b)

CHAPTER 10- CONCLUSION

We applied the Hough transform in overlapping time-slices of the TFR images to increase the robustness against noise and interference terms, present in the TFRs. Moreover, by exploiting the instantaneous frequencies estimates obtained for many spatial frequency pairs using FCP, we increased the robustness and accuracy of our method. Having estimated accurately the objects velocities, the proposed objects segmentation approaches produced correct and meaningful results. The experiments on various synthetic and natural sequences verified the effectiveness of the proposed method.

By applying Space-locally using 2-D Fourier Transforms.

It Increased Computational effort, compared to constant motions approaches and also benefit of increased accuracy.

Our future research on achieving real-time motion tracking characteristics, for both Monochorome and Color image sequences.

REFERENCES

1. Y.Wang, J. Ostermann, and Y. Zhang, Video Processing and Communications.Englewood Cliffs, NJ: Prentice-Hall, 2002.

2. B. K. B. Horn and B. G. Schunck, "Determining optical flow," Artif.Intell., vol. 17, pp. 185–203, 1981.

3. M. J. Black and P. Anandan, "The robust estimation of multiple motions: Parametric and piecewise-smooth flow-fields," Comput. Vis.Image Understand., vol. 63, pp. 75–104, Jan. 1996.

4. E. H. Adelson and H. R. Bergen, "Spatiotemporal energy models for the perception of motion," J. Opt Soc. Amer. A, vol. 2, pp. 284–299,1985.

5. B. A. Watson and A. J. Ahumada, "Model of human visual-motion sensing," J. Opt Soc. Amer. A, vol. 2, pp. 322–342, 1985.

6. C. Mota, I. Stuke, T. Aach, and E. Barth, "Divide-and-conquer strategies for estimating multiple transparent motions," in Lecture Notes on Computer Science. Berlin, Germany: Springer, 2005, vol. 3417.

Publisher: Eliva Press SRL

Email: info@elivapress.com

www.ingramcontent.com/pod-product-compliance
Lightning Source LLC
Chambersburg PA
CBHW071244220526
45468CB00002B/996